discovermore

Exploring Physical Science

Energy

Marie Harts

Britannica
Educational Publishing
IN ASSOCIATION WITH
THE ROSEN PUBLISHING GROUP

Published in 2024 by The Rosen Publishing Group, Inc.
2544 Clinton Street, Buffalo, NY 14224

Editor: Shannon Harts
Book Design: Rachel Rising

Photo Credits: Cover zieusin/Shutterstock.com; (series background) Dai Yim/Shutterstock.com; p. 4 Sharomka/Shutterstock.com; p. 5 Kampan/Shutterstock.com; p. 6 ElenaNoeva/Shutterstock.com; p. 7 Surasak_Photo/Shutterstock.com; p. 8 Art by Dr. Dan/Shutterstock.com; p. 9 Juris Teivans/Shutterstock.com; p. 10 Vladimir Batishchev/Shutterstock.com; p. 11 Eduard Goricev/Shutterstock.com; p. 13 Patrick Ziegler/Shutterstock.com; p. 13 MEE KO DONG/Shutterstock.com; pp. 14, 25 VectorMine/Shutterstock.com; p. 15 Montypeter/Shutterstock.com; p. 16 Viktoriia Hnatiuk/Shutterstock.com; pp. 17, 26 Africa Studio/Shutterstock.com; p. 18 Martin Mecnarowski/Shutterstock.com; p. 19 ABCDstock/Shutterstock.com; p. 20 Triff/Shutterstock.com; p. 21 Nadezda Murmakova/Shutterstock.com; p. 22 Aleksandar Malivuk/Shutterstock.com; p. 23 Followtheflow/Shutterstock.com; p. 25 James Marvin Phelps/Shutterstock.com; p. 27 Breedfoto/Shutterstock.com; p. 28 Dean Drobot/Shutterstock.com; p. 29 Rudmer Zwerver/Shutterstock.com.

Library of Congress Cataloging-in-Publication Data

Names: Harts, Marie, author.
Title: Energy / Marie Harts.
Description: Buffalo : Britannica Educational Publishing, [2024] | Series: Discover more: exploring physical science | Includes index.
Identifiers: LCCN 2023027385 (print) | LCCN 2023027386 (ebook) | ISBN 9781641900454 (library binding) | ISBN 9781641900447 (paperback) | ISBN 9781641900461 (ebook)
Subjects: LCSH: Power resources--Juvenile literature. | Force and energy--Juvenile literature.
Classification: LCC TJ163.23 .H375 2024 (print) | LCC TJ163.23 (ebook) | DDC 621.042--dc23/eng/20230614
LC record available at https://lccn.loc.gov/2023027385
LC ebook record available at https://lccn.loc.gov/2023027386

Manufactured in the United States of America

CPSIA Compliance Information: Batch #CWBRIT24. For further information contact Rosen Publishing at 1-800-237-9932.

Contents

What Is Energy?

Energy is power used for many purposes. Energy makes objects move. It also makes objects warm. Energy causes machines to work and living things to grow.

Machines can turn energy from nature, such as wind and sunlight, into electricity. Your body can turn food into energy that powers your movements, such as running!

People use volts to measure the potential energy in electricity.

Energy comes in many forms. People use energy from many different sources. They use the power of wind, water, and sunlight. Plants and animals provide people with energy in the form of food. People also burn oil, coal, and natural gas for energy. They get nuclear energy from uranium.

Each of these forms of energy can be described as either potential energy or kinetic energy. Based on the law of the conservation of energy, it's impossible to create or destroy energy. Instead, energy changes form.

Consider This

Electricity and natural gas are the most common energy sources used in U.S. homes. What machines in your home may use electricity or natural gas?

The Energy of Motion

The energy of motion is called kinetic energy. The word "kinetic" means "active" and relates to the motion of physical objects. All moving objects have kinetic energy, even atoms. Atoms are the tiny building blocks that make up everything.

When a roller coaster speeds downwards, it converts potential energy to kinetic energy.

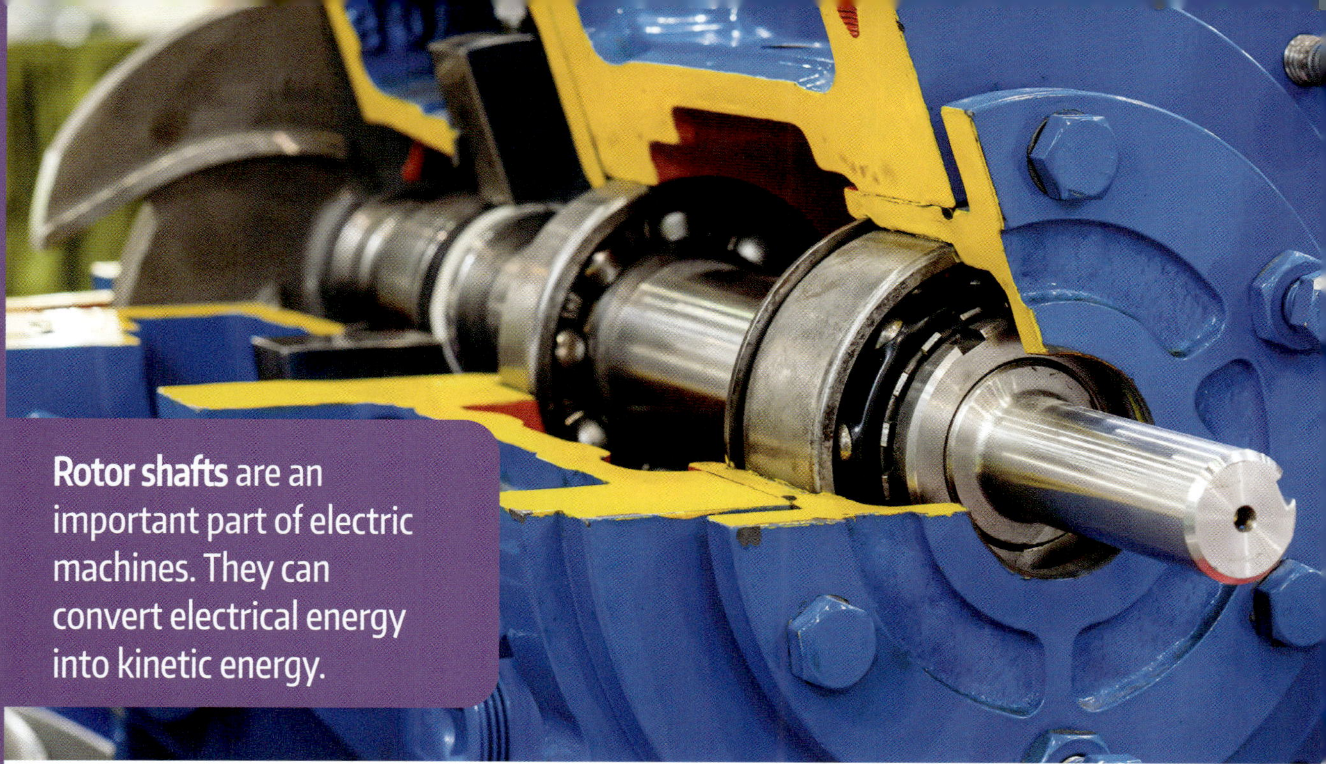

Rotor shafts are an important part of electric machines. They can convert electrical energy into kinetic energy.

Forms of kinetic energy perform work. The work done by kinetic energy can be seen everywhere. River water has kinetic energy that can dig deep canyons as a river rushes downhill, pulled by gravity. Sailboats use the kinetic energy of wind to move along. Wind turbines and other ways to make electricity use a machine called a generator. In a generator, water, air, steam, or gases spin blades on a **rotor shaft** to make electricity.

WORD WISE

A rotor shaft is a long, cylindrical part, often a rod or tube. It transmits, or carries, power created by a rotor.

Stored Energy

Energy that is stored, or conserved, is called potential energy. An object with potential energy has the ability, or potential, to move. Potential energy comes from the position, shape, or state of an object.

An inflated balloon's stretched rubber has potential energy.

The springs on a trampoline have potential energy. You also have potential energy before you jump!

Sometimes people store energy on purpose. Charging a **battery**, stretching a rubber band, and pushing down on a metal spring are all ways of storing energy. Sometimes energy builds up gradually. For example, a tree grows for many years before people cut it down and burn it for light and heat. Potential energy is latent, or ready to be released, such as a rock perched high atop a cliff's edge. When the rock falls or is thrown off the cliff, its potential energy becomes kinetic energy.

WORD WISE

A battery is a device that can change chemical energy stored inside its materials into electricity. It is often placed inside other devices and machines to supply them with electricity.

Combining Energy Types

When you combine potential and kinetic energy, you get mechanical energy. Mechanical energy is the energy an object has because of its motion and position.

A pendulum, shown above, demonstrates mechanical energy. When the pendulum's swinging object reaches its highest point, its energy is potential. When the object swings down, its energy is kinetic.

Shooting a bow and arrow is also an example of mechanical energy. The arrow drawn back in the bow has potential energy. The arrow has kinetic energy when released.

Machines use mechanical energy to do work. A hammer uses mechanical energy to drive nails into a board. When the hammer is raised, it has potential energy from the work done to lift it. When the hammer is moved toward a nail, potential energy becomes kinetic energy, which does the work of driving the nail into a board. The hammer's kinetic energy is transferred to the nail each time the nail is struck.

Bowling is another example of mechanical energy. A bowling ball has potential energy when held. This energy becomes kinetic when it rolls down the lane!

Consider This

A pendulum is an object that hangs from a fixed point so it can swing back and forth. Can you think of machines that use pendulums and mechanical energy? One example is a wrecking ball!

Transforming Energy

Although energy can't be created or destroyed, it can be transformed. This means it can change forms. Energy is often changed into a form that meets the needs of a plant, person, or animal.

A plant such as a tree changes energy from sunlight into chemical energy. When an animal eats a plant, it changes this chemical energy into kinetic energy when it moves. A car or truck changes chemical energy into mechanical and heat energy. A nuclear reaction changes nuclear energy to **radiant** and heat energy. A lamp changes electrical energy to radiant and heat energy. Throughout history, people have created tools to change and use energy.

WORDS TO KNOW

Radiant means sent out from a source in rays or waves that cannot be seen.

Plants can change the sun's energy, water, and the gas carbon dioxide into oxygen and energy in the form of sugar. This is called photosynthesis.

compare and contrast

Do you think some forms of energy are easier to change than others? Electricity, for example, can be changed into heat and light quite easily. However, changing thermal, or heat, energy back into electricity is not as easy.

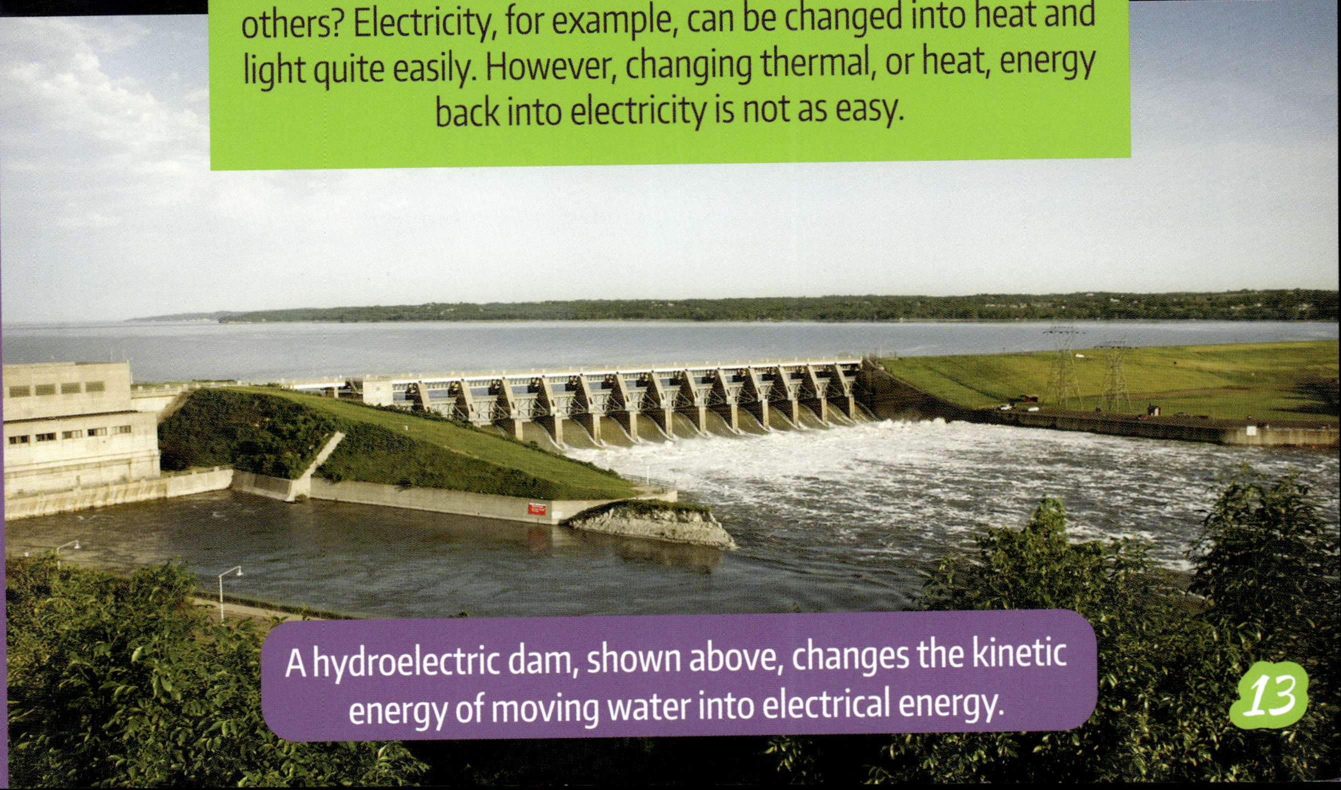

A hydroelectric dam, shown above, changes the kinetic energy of moving water into electrical energy.

Feel the Heat

When the atoms that make up a substance vibrate, or quickly move back and forth, they create heat energy. This increases the **temperature** of the substance. Heat energy can be used on purpose, such as to make stoves hot for cooking or to let matches start a crackling fire. It can be dangerous, or unsafe, if heat energy starts a fire where people don't expect it.

Wells sometimes as deep as several miles are built to reach ground that is hot enough to create electricity from geothermal energy. Turbines and generators help turn the hot water or steam produced into electrical energy.

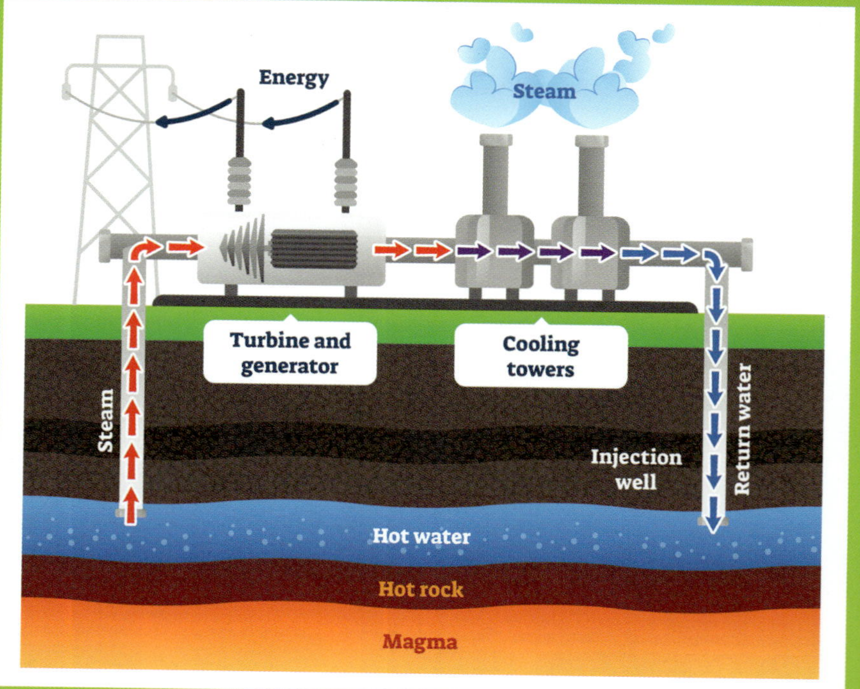

An electric stove often has coils or rods under its surface. Electricity heats up these coils or rods, allowing them to cook food.

For many years, people have used hot water and steam from inside the Earth for heating. Today, in some regions, or areas, hot water or steam is run through pipes to people's homes to provide heating and cooling. This type of energy, called geothermal energy, can also be used to create electricity.

WORD WISE

Temperature is how hot or cold something is.

Mixing Atoms

Mixing atoms can cause some to form new bonds. This is a chemical reaction. Chemical reactions can use energy or they can release it. Often the energy is in the form of heat.

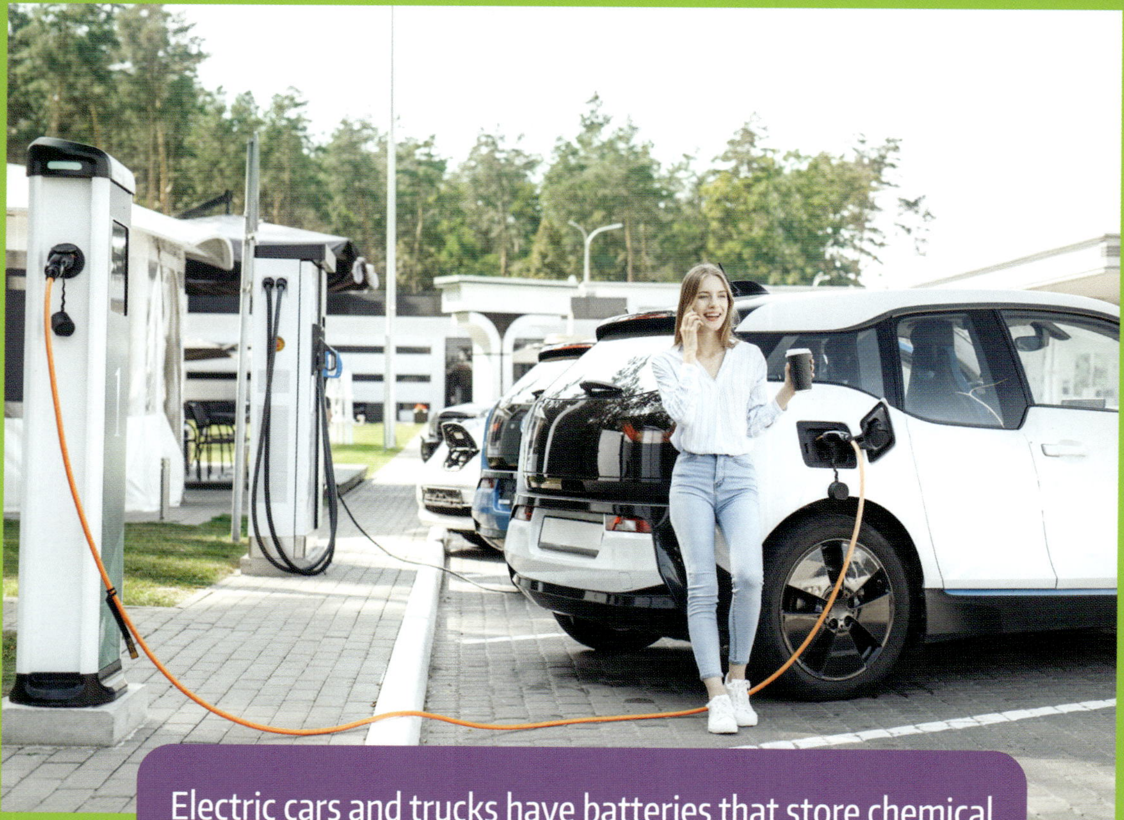

Electric cars and trucks have batteries that store chemical energy and can change this chemical energy into electricity.

When you eat, you release chemical energy stored in food. Your body can turn the chemical energy into mechanical energy for everyday activities.

Plants use energy from sunlight to bind together atoms from air and water so that they can grow. Sunlight energy is used in the chemical reaction, but it doesn't just disappear. It changes into stored chemical energy in the body of plants. When animals eat plants, they turn the chemical energy into energy they use to live and move. When fire burns wood or oil, that's another chemical reaction. The carbon in wood combusts due to the presence of oxygen, forming carbon dioxide and water.

Consider This

Why do you think batteries might be important for energy sources such as wind and solar? One reason is that they can store electricity from the wind and sun in the form of chemical energy when the sun isn't shining and the wind isn't blowing.

It's Electric!

Atoms have tiny particles called electrons around their nucleus, or core. Electricity is created when electrons move between atoms. In nature, electricity can be seen in bolts of lightning. Lightning is a large number of electrons flowing through the air all at once. Lightning releases huge amounts of energy.

Scientists have learned how to generate, or make, electricity from other forms of energy. Electricity is useful because it can be sent through wires to where it is needed. It powers heaters, light bulbs, and computers. Electricity can be made from **renewable energy** sources and from non-renewable energy sources, such as fossil fuels.

In 2020, wind power was the most commonly-used form of renewable energy in the United States.

Transmission lines have wires that carry the electricity used in an electric power system. They can carry a lot of power over large areas!

A lot of electricity in the United States is made by burning fossil fuels. This produces gases that trap heat around Earth, causing climate change.

WORD WISE

Renewable energy is energy from sources that are naturally replenished, or renewed, such as sunlight and wind.

Big Energy, Small Source

Nuclear energy comes from the very tiny nucleus of an atom, but it can create a lot of power! When nuclear energy is released, it changes into other forms, including heat and light. Nuclear energy can be released through nuclear fission, when the nucleus of an atom splits into two or more parts, or through fusion, when two nuclei join to form a single nucleus.

Nuclear fusion creates energy for stars, including our sun. In 2022, for the first time, scientists created a fusion reaction that produced more energy than was needed to create it.

A nuclear power plant changes the heat and light from nuclear fission into electricity. There are about 92 in the United States, as of 2022.

Nuclear energy was first used in powerful bombs that the United States dropped on Hiroshima and Nagasaki, Japan, during World War II. Nuclear power plants heat water to make steam and turn turbines, creating electricity. Today nuclear power plants that create electricity use nuclear fission.

Consider This

Nuclear power plants create nuclear waste, which can be dangerous and hard to get rid of. However, nuclear power creates electricity constantly and doesn't pollute the air like burning coal, oil, and natural gas can. Why do you think that is?

Super Solar

The sun provides energy for every living creature on Earth. At the sun's core, great pressure squeezes hydrogen gas, causing a nuclear reaction. Atoms of hydrogen join to make helium—and lots of energy! That energy radiates into space as sunlight.

In 2021, solar power created 3 percent of all U.S. electricity.

Homes can be designed, or planned, to save energy for heating by using passive solar energy. One way is to have south-facing windows.

On Earth, the sun's light warms the air and causes the wind to move. Wind helps carry that heat and energy around the world. Water evaporates when solar energy heats lakes and oceans. People use solar energy in many different ways. **Solar panels** change radiant energy into electricity. Passive solar energy projects use sunlight more directly. Greenhouses and sunrooms use passive solar energy as they are heated directly by the sun's rays passing through windows.

WORD WISE

Solar panels are devices that can change light from the sun into electricity. They do this by absorbing, or taking in, particles of energy called photons. This creates an electric current.

Digging Up Energy

People have used coal from deep underground for thousands of years to create heat. But there are other fuels as well. When plants and ocean plankton die, they decay, or break down. Long ago, some remains became deeply buried and underwent a slow chemical reaction over millions of years to create what are called fossil fuels. **Petroleum**, natural gas, and coal are all examples of fossil fuels.

Fossil fuels store chemical energy. These fuels burn in a fast chemical reaction. They also burn heat energy very quickly. The reaction makes smoke, soot, and harmful gases. This air pollution can lead to health conditions like asthma, which affects the lung airways.

WORD WISE

Petroleum refers to oil that comes from rocks or the earth.

Fossil fuels are formed from fossils. These are the remains or traces of creatures that lived millions of years ago, transformed by heat and pressure.

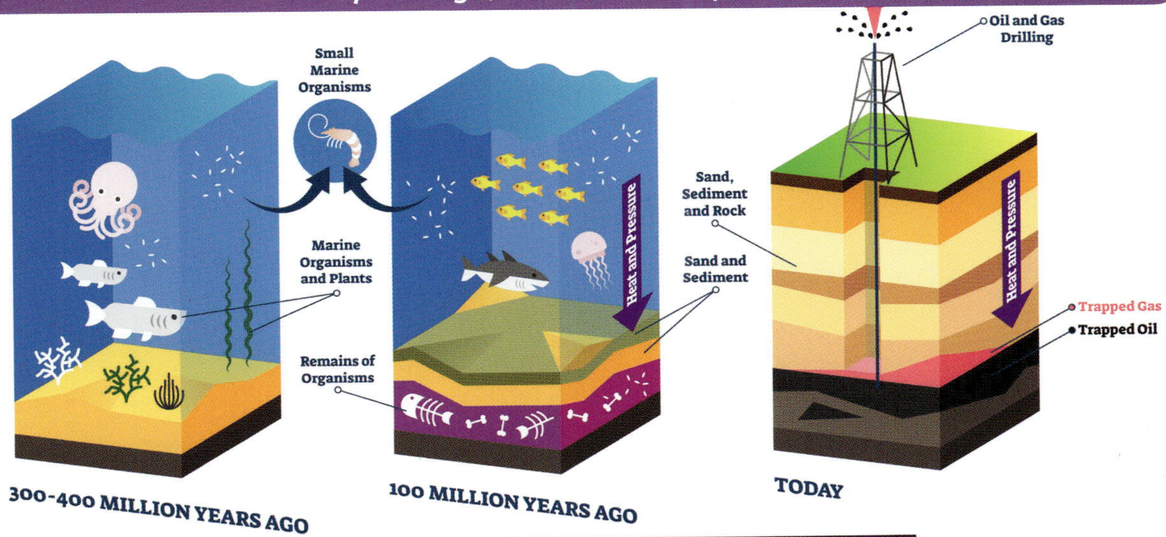

compareandcontrast

Both solar energy and fossil fuels can create electricity. Fossil fuels are the remains of plants and animals that got their energy from the sun. However, this energy is stored in a secondary form.

Many U.S. coal plants, such as the Navajo Generating Station shown here, have closed as clean energy such as solar becomes cheaper.

The Flipside to Fossil Fuels

People are worried about how fossil fuels cause climate change and air pollution. They are also being used up much too quickly. Scientists are in search of other sources of energy to replace fossil fuels. Some alternative energy sources have been used for many years. These include sunlight, wind, rivers and oceans, and heat from inside Earth. Some biofuels are made from plants, such as soybeans, or from organisms called **algae**.

Biofuels are made from recently living plant matter. They are more renewable than fossil fuels because they are made from crops that can be replenished, or replaced, faster.

Tidal energy is another alternative energy source. Electricity is created from the ocean forces that cause tides.

Most alternative energy sources are renewable. Unlike fossil fuels, many alternative energy sources are not likely to run out. For this and other reasons, the use of renewable energy sources has been growing in the United States—in fact, the use of renewable energy hit a record high in 2021.

WORD WISE

Algae are small, plantlike creatures that can turn sunlight into energy. They can also be used as a renewable energy source because some types have oils that can fuel planes, cars, trucks, and trains.

Energy All Around

You can find energy sources all around you! The sun radiates energy out into space. Energy from the sun's light mixes Earth's air and water. Solar energy changes form as wind and water move around the world. Even Earth's core is warm, from gravity pressure and uranium radiation. There is plenty of potential energy and kinetic energy changing form all the time.

Using electricity more efficiently can prevent pollution from burning fossil fuels to make energy. Having computers enter a "sleep" mode when not in use is one way to do this.

Biogas, energy that comes from the waste of animals like cows, is a renewable energy. After some clean-up, it can be a replacement for natural gas.

All that energy makes Earth a good place for all kinds of living things. People can use energy in ways that are renewable. However, people can also use energy in ways that create pollution and waste. People and businesses are switching to using more renewable energy to help keep the planet clean and healthy for all creatures.

Consider This

Energy for electricity and heating can even come from human and animal poop! This energy from the waste being broken down is called biogas! Can you think of other unexpected sources of energy?

Glossary

canyon: A narrow valley that is deep with steep sides.

chemical: Matter that can be mixed with other matter to cause changes.

climate change: Long-term change in Earth's climate caused mainly by human activities such as burning oil and natural gas.

combust: To burn.

convert: To cause to change form. Conversion is the act or process of converting.

cylindrical: Shaped like a cylinder, or a shape that has circular ends and straight sides.

efficiency: The quality of being efficient, or able to produce a desired result with little to no waste.

evaporate: To change from a liquid to a gas.

fossil fuel: Matter formed over millions of years from plant and animal remains that is burned for power.

gravity: The force that pulls objects toward the center of a planet or star.

oxygen: A colorless, odorless gas that many animals, including people, need to breathe.

plankton: A tiny plant or animal that floats in the ocean.

potential: Existing as something that is possible and capable of happening.

pressure: A force that pushes on something else.

radiant energy: Energy that radiates from a source in waves.

radiation: Waves of energy that come from a nuclear reaction and that can be dangerous in large amounts.

source: The cause or starting point of something.

uranium: An element that is used to make nuclear energy. It is radioactive, which means it emits, or gives off, an unsafe energy called radiation.

For More Information

Books

Dickmann, Nancy. *Energy*. Tucson, AZ: Brown Bear Books, 2023.

Fabrizio, Alex. *Green Energy*. Elmhurst, NY: Starry Forest Books, Inc., 2021.

Pettiford, Rebecca. *Energy*. Minneapolis, MN: Bellwether Media, 2019.

Websites

Can We Produce Energy by Exercising?
www.wonderopolis.org/wonder/Can-We-Produce-Energy-by-Exercising
Learn about how your movements could be used as an alternative energy!

Energy
kids.britannica.com/kids/article/energy/353100
Check out more information about energy from Britannica on this webpage for kids.

Renewable Energy
education.nationalgeographic.org/resource/renewable-energy/
Learn more about renewable energy with National Geographic.

Publisher's note to educators and parents: Our editors have carefully reviewed these websites to ensure that they are suitable for students. Many websites change frequently, however, and we cannot guarantee that a site's future contents will continue to meet our high standards of quality and educational value. Be advised that students should be closely supervised whenever they access the internet.

Index